FARM MACHINES

KATHY MIDDLETON

CROP SPRAYER

A Crabtree Roots Book

Crabtree Publishing
crabtreebooks.com

School-to-Home Support for Caregivers and Teachers

This book helps children grow by letting them practice reading. Here are a few guiding questions to help readers build comprehension skills. Possible answers appear here in red.

Before Reading:

- What do I think this book is about?
    - *I think this book is about crop sprayers.*
    - *I think this book will talk about what crop sprayers do.*
- What do I want to learn about this topic?
    - *I want to learn about what jobs crop sprayers do.*
    - *I want to learn why crop sprayers are important farm machines.*

During Reading:

- I wonder why...
    - *I wonder why farmers use a crop sprayer.*
    - *I wonder why crop sprayers are needed on the farm.*
- What have I learned so far?
    - *I have learned that crop sprayers add fertilizer to soil.*
    - *I have learned that crop sprayers help get rid of insects.*

After Reading:

- What details did I learn about this topic?
    - *I have learned that some crop sprayers help water plants.*
    - *I learned that some crop sprayers attach to tractors.*
- Read the book again and look for the vocabulary words.
    - *I see the words **crop sprayer** on page 3 and the word **farm** on page 12. The other vocabulary words are found on page 14.*

This is a **crop sprayer**.

A crop sprayer adds **fertilizer** to soil and sprays fields for weeds and insects.

Some crop sprayers are **vehicles**.

Some crop sprayers attach to **tractors**.

Water sprayers give crops water so they can grow.

Crop sprayers
help plants grow
on a **farm**.

Word List

Sight Words

a
adds
and
are
attach
can
crops
do
for
give
grow
insects
is
jobs
many
on
plants
so
soil
some
sprays
they
this
to
water

Words to Know

crop sprayer **farm**

fertilizer **tractors** **vehicles**

47 Words

This is a **crop sprayer**.

A crop sprayer adds **fertilizer** to soil and sprays fields for weeds and insects.

Some crop sprayers are **vehicles**.

Some crop sprayers attach to **tractors**.

Water sprayers give crops water so they can grow.

Crop sprayers help plants grow on a **farm**.

FARM MACHINES

CROP SPRAYER

Written by: Kathy Middleton
Designed by: Under the Oaks Media
Proofreader: Crystal Sikkens
Production manager: Candice Campbell
Prepress technician: Katherine Kantor

Photographs: All images from Sutterstock

Crabtree Publishing

crabtreebooks.com 800-387-7650

 In Canada: We acknowledge the financial support of the Government of Canada through the Canada Book Fund for our publishing activities.

Printed in the U.S.A./CP2025

Published in Canada
Crabtree Publishing
616 Welland Ave.
St. Catharines, Ontario
L2M 5V6

Published in the United States
Crabtree Publishing
347 Fifth Ave
Suite 1402-145
New York, NY 10016

Library and Archives Canada Cataloguing in Publication
Available at Library and Archives Canada

Library of Congress Cataloging-in-Publication Data
Available at the Library of Congress

Hardcover: 978-1-0398-8107-5
Paperback: 978-1-0398-8468-7
Ebook (pdf): 978-1-0398-8227-0
Epub: 978-1-0398-8347-5
Audio: 978-1-0398-8510-3
Read-Along: 978-1-0398-8552-3